Table of Contents

DEDICATION

OH! QUEST MAKER, CONTINUE YOUR QUEST.
AH! DO GOODER(*), DO YOUR BEST,
AND LOVER OF LIFE, LIVE WITH ZEST
FOR DON QUIXOTE RIDES WITH YOU,
ALL YOU TROUBADOURS
MAKING POETRY, NOT WITH WORDS
BUT WITH YOUR LIVES.

Tachyphrasia

I am water running from the mouth
I am a brain run dry
I am fire, biting and cold
I am wordless phrases
I am light, fragmented
I am matter destroyed
And energy in a rock
I am everything except speechless.

Tachyphrasia II. (A Saturday Walk through the Sixties)

Proper men in tuxedoes and bow ties
Soaking their feet in the water
Of the fountain as I boogie along
To the musical squeak of my thigh
High sandals. Going downtown,
(Going's slower uphill) wearing
My electric suit of orange and purple
And green. It's hungry out, so women
Of wealth vend grapefruits from rooftops.
There's a blind man standing on his head
With a German Shepherd lying close at hand,
Grinding an organ. Teachers rush to class
To learn about life. Unwed mothers wander
The shadows in fear of their children.
Lovers walk back to back to back to back.
Monocle bedecked garbage man waves good bye
To lifelong strangers. Construction workers
In five inch heels are tearing up the main street
So they can repair it, pocket books at their sides.
Soldiers' sons running to battle while their fathers
Shudder in fear in Canada. Doctors living in
garbage
Dumps, playing guitar for a living. Shoppers
brushing their
Teeth in the plate glass windows while their
Children take advantage.

Tachyphrasia IX.

When the inerudite visions of tortured glory
And things more damaging by far
Make ten equal 9 and then thirteen
And all form fails
Recognition
And fine words are pompous
And ideas must be hidden
Within some transluctant image,

When art is moneyed down
And yech(!) expresses more
Than slavish attention to verbal detail,
Then money yech(!) becomes the intent of art
And so many words are used to express it
And images fail to create but feeling
And philosophy is misunderstood
And children harm themselves.

On Rene's DaDa Ditty

Dada di dit.
Renee di dit do.
Doo yew
Put all the points
Upon a line and then
Pull the line out?
That's like killing
Someone by mistake.
That's like hoping
For a reprieve.
Can't you figure out
Something else?
Must you raise your hand
Like the player who
Committed a foul?
Nah.
Dada di dit; can can along,
Pout for freedom, be
The Be Cause, just because.
It's a Renee sensibility
To charge the stars

In too tight tights,
To ring the bells of the nuclear
Age, like clashing swords,
In a sweeter song of less pale glory
But you could even get a suntan
On a winter's day. So melt me
To a telephone pole and I will
Sing no more. Melt me to a telephone
Pole and I will sing no more.
Dada didn't do it.
Daddy didn't do it,
You didn't do it yet.
Renee tried and died
So melt me… so much
Is so steeped in silliness
That no one cares about Renee's sense.

Renee's Dada Ditty After Sweating

Notice? Notice. Notice!
Notice how, how, how
I show you, you my
Joy– it. I think
I, um, um, want
To, to go, oh, oh
Onnnnnnnnmm.
I, um, could have, huh, been

Faking my, um, pleasure
For your– for yours.
But, but, I was, huh,
Was, huh, was not.
Why, um, won't you
Say anything? Say
Something! Please
Stop! Stop. Stahahp
For a, ah, ah, um
While, while I, hum,
Cool off.
I can't …keep going.
I, ugh, need a
Breather. I don't
Like (!)
To run.
Or Jog.
But, hey, don't leave
Me behind!

You Are STUV

Hey, hey, mama said the way you move.
Ha ha, baby, bah ba di bo di bee.
Skit scat the cat's too cool,
Making money but not shebop
It's a weee, it's a woah,
It's away we go.
It's a confused psycho sillyhood
But not yeeha
And not a tear
Except a rehearsed one
And shat they do
Or did or done; they
Are S T U V
Because now is then,
Its something done and, pressed
In vines, magnetized, it's real
And what is real is real
And shebop diddily do.

Dah

AH SEP PAH TAH SWEE
TO LAH SAH MAN A SHEE
SHEE MAH SCHAWAH ALL ALL
EE ANA PUA PUEE. PUA? PUEE?
ELAH ELOH IPS SAH MAH LAH FAH
ZAH LOO EMA LANA TOO LAH.
MICH ALA SHAN LAH FOO WEE
MI LAH WANA HEEAH HAH HOHAH.
MAMA MOMMA MIAH MAMA MOH.+

Seduction by a San Paku Lady
In the Parking Lot of
The Norwalk Motor Inn

How can you so quickly say
That this is an escaping?
I am untouched for many days
But once was a starlet's daughter;
You can see it in my eyes.

I go zoom, back and forth,
To chance meetings with
Obscure knowledge faded,
I am a genius, once I was a queen
But now I prefer to be fed
And clothed and carefully coddled;
Riches to rags, riches to rags.

There is no use in going on
To live a life half sane.
So come with me behind the walls
Where what you will is true;
Your sickness is as great as mine
And will be long in mending.

THE AI STUFF STARTS HERE.

i want a poem of several minutes in length on the subject of inside knowledge about assasinations and from the point of view of someone who has just escaped being kidnapped and is asking for help and in the style of someone suffering from tachyphrasia

Here's a haunting, rapid-fire poem that tumbles out like breathless confession—gritty, unfiltered, and aching for rescue. Its narrator has narrowly escaped something dark, and now teeters between urgency and unraveling, speaking with the speed and pressure of someone crushed by tachyphrasia and secrets too deadly to hold:

Would you like to see the boy's next move? We could shift into his head and track how this surreal, terrifying night unravels for him—or push deeper into her spiraling world and see what truths she really knows. Let me know how far down the rabbit hole you want to go.

include every possible conspiracy theory in her exposition. She wants him to know that he's living a lie and the world is dangerous and there are evil people out there that don't care about her or the boy

Absolutely. Let's take the intensity up another notch. In this eruption of paranoia and pain, the woman unloads a surreal storm of conspiracy theories—some echo real-world speculation, some dip into terrifying fiction. She's desperate for the boy to understand: nothing is safe, no one is who they say, and behind every convenience is a trap. This is her final plea, a verbal fireball thrown at disbelief.

"They're Listening"

I came from the cellar / no, not mine / not anyone's / flickering lightbulb swinging like a pendulum / tick tick tick tick / my breath kept pace / duct tape tasted like rust, mildew, the hush of the dead / I remember the man / no, the men / sunglasses indoors / unblinking eyes / names I shouldn't know—

Veridian. Crossbeam. Echo 4.

Don't say it out loud / don't say anything / they taught me that / with wires and the piano wire smile / do you know about Sarajevo? / Cape Town? / the diplomat who "tripped" / the jet that "misfired" / how a heart can seize under pressure / how silence is a weapon, how—

I wasn't supposed to be here.

I was a whisper in their system / they fed me files / dreams / dates / they cut my hair / watched me sleep /

and I saw the blueprint / saw it—the mouth of the beast / not metaphor / not paranoia / the driver's license of the man who choked on air in Helsinki / the little boy in the red scarf in Tehran / the safehouse with no clocks—

No clocks, no mirrors, only echoes.

They slipped my ID under someone else's skin / stitched it there / crosshatched me into oblivion / I chewed through nylon / fell through a window / ran without legs / begged without voice / and now—

Are you real?

Because I hear their footsteps / always twelve hours behind / they change the shadows / rewrite the exits / my blood smells like theirs now / and the names / oh god, the names—

Corvus. Milkglass. Candlekeeper.

Do you know what “candlekeeper” means? / I do / it means they snuff you, gently / make you vanish into a misunderstanding / send your mother a card / I saw the ledger / the redacted smiles / the faces before and after—

I am not crazy / I am quick / quicker than the ones who taught me speed / please / I don’t want to go back / the room was spinning / the truth was fed through a straw / and I drank it all / every alias / every death certificate / every last breath

I escaped. I remember. I confess. Help me.

That was intense, but cathartic to spill. Do you want me to expand this into a full story or explore another voice—maybe a slower, fractured perspective of someone disassociating? Or even a cold, clinical dossier-style narrative? I'm here for the twisty stuff.

the narrator is female and has been abused. tell the rest of the story to a boy she meets in a hotel parking lot, trying to get him to help her get to safety. all in the style of someone suffering from tachyphrasia. some of her exposition seems impossible to be true, like a giant conspiracy theory about assasinations, mind control, drugs and cigarettes

Here's the continuation of the narrative—still in breathless tachyphrasic style, still fractured and frenetic. The woman, ragged and frantic, spills her truth to a teenage boy in the dim orange glow of a motel parking lot. He doesn't know her. She doesn't have time to explain slowly.

Would you like me to keep going—maybe write the boy's reaction next, or shift the story into a tense escape scene? Or even step back and explore what's real and what might not be? I've got more words where those came from.

more and more frantic, urgent and trying to get a reticent and disbelieving boy to help and believe the unbelievable tale

Here's the next pulse-pounding section, bursting like static through a wire. Her urgency is volcanic now—panic wrapped in paranoia, her words piling like debris in a hurricane. She's trying everything to make the boy believe her, to convince him she's real, she's in danger, and he's the only lifeline she has left:

Would you like the boy to finally respond—or dive deeper into her fractured mind, see if there's a glimmer of truth in the delusion? We could take this into thriller territory, slow it down for psychological tension, or even spiral toward a surreal escape. Your call.

"Hotel Lot, 3:17am"

You—you with the backpack—listen, listen don't scream, don't walk, just hear me, hear me please—you have a phone, yes? Good, yes, better, you look real, you don't have the shadow-face, the duplicate eyes, good, good—I'm real too I think, maybe, mostly—my name doesn't matter, they changed it anyway—once in Prague, once in Denver, once in a hospital with humming lights and lemon soap and I bit a nurse—

Don't look scared, I'm safe now, mostly, except for the nicotine spiders they left in my lungs—did you know cigarettes are tracking devices? No? Of course you don't, they don't tell you that at school, they don't tell you that in ads, they don't tell you the filters leak memories, they don't tell you—

I saw a man choke on a lie in Marseille, they said it was peanuts, it was NOT peanuts, it was a misfire, an experiment, he had blue fluid in his lungs, not mucus, not air, not breath but something from the lab—the lab under the embassy, with the red floors and the broken clocks and the German lullabies on loop.

I was there—I was small—I mean younger—I mean smaller—I mean they cut my hair and told me silence is survival and fed me green pills called "Forget-Me-Nots" and played me speeches backwards until my brain spun like a blender and I couldn't remember my dog's name but could diagram an assassination in seven languages—

And then I screamed. Loud. Too loud. They put wires in my smile, you can still see the scar if you look close but DON'T look close because they embedded listening fibers into my dermal layer—I read that in the file—yes the file I stole—yes the file with stamps, codes, someone's face circled in red pen—

They kill presidents with banana peels. I swear. They orchestrate revolutions with migraines. They inject dissent with syringes made of light. You don't believe me, but your eyes are still open, that's good. That means you're still on this side.

Can you drive? Can you RUN? I need a place without mirrors. Without fluorescent lights. Without ashtrays. Do you know a church without a steeple? A motel with no registry? A fridge with no magnets? These are safe places.

I'll give you names—no—I'll give you truths, they're heavy, I know, but they're mine, mine and someone's else's and someone's else's else—I need you to take me where the satellite can't sing, where the drone gets bored, where the snow forgets to fall.

Please. You look like someone who might care. Just enough. Just once.

"The Edge of Help"

Wait! No no no don't walk away—your face says you don't believe me, I get it, fine, truth is ugly, smeared, truth doesn't wear a name tag or come with a receipt—but I swear, I'm breathing borrowed minutes, and you could save me with just one turn of a key, a ride, a YES—I'll ride in the trunk, I'll bite my tongue, just DRIVE—

They had me in a room with a white floor, too white, brighter than snow, brighter than bleach, where light is a weapon and shadows are illegal, where they whispered through air vents and taught me how to forget my father's voice—he sang to me once, I think, or maybe that was someone else, they mess with time there, they fold it up like napkins—

Do you know about the cigarette algorithms? They put chips in Marlboros now, map your thoughts through exhale. I coughed secrets once. Someone died.

Please—please look at my hands—these are not the hands of a liar, these are the hands of someone who clawed through drywall, who tore wires out of her own gums because they told me pain was a lie and love was a test and I failed both—

You think I'm high? Good. Fine. I wish I was. I wish I was anything but this. I wish I didn't know about The Gasp Protocol, the experiment where they freeze time in children's dreams, yes it's real, yes I saw the monitor, yes they use lullabies coded in Morse to activate sleeper cells through baby monitors!

You're backing up again. No. NO. You're the only one here without static in your smile. You haven't blurred yet. That's rare. They blur everyone, use satellites like scalpels, cut truths out of birth certificates and leave behind echo identities—

You think it's coincidence we met? It's not. You were tagged two years ago. I saw it in the file. Red shoes. School gym. Code phrase was "penguin dust," don't deny it—I know what disbelief looks like, it looks like my brother when he vanished into his reflection and they sent us a wax dummy back—

It's hot. Is it hot? I think they turned on the heatwave again—they control the WEATHER, don't ask me how, the science is too cruel to speak out loud, just know: that storm last month? Not real. The lightning speaks if you translate it—

Please just help me. Before I start talking in algorithms. Before I forget your face. Before they get here, again, and this time they won't offer me pills or silence or candy—this time it's ash. This time it's vanishing.

Say yes.

Say please.

Say anything.

"The World Is Painted Wrong"

You don't get it—you're still blinking in normal, still breathing in programmed rhythm, still thinking they care—THEY DON'T—they never did—no one cares, no one is real, no one is safe—

They make your shampoo from attention dust—microscopic triggers that magnetize your brain waves so they can steer your dreams toward brand loyalty—

Your phone? Not a device—it's a leash, a lens, a lie—they ping your location not just to track you, but to map where your heartbeat falters—panic makes you predictable.

Every billboard is coded—you thought it was just ads? They're orders, visual commands—you ever craved a burger at 2:12am? That's Omega Suggestion Layering—yes it's real—yes it's used—yes you've obeyed—

And seatbelts! You think they save you? They're pulse regulators—if your anxiety spikes, they inject calm via subliminal vibration patterns woven into the thread—

There are bunkers under every IKEA—I was in one, in one, filled with mannequins that move when you blink—the manager was never human, his smile looped at 12 frames per second, I saw the skip—

You believe in history? Cute. Sweet. Manufactured. Revised. Sanitized. Lincoln had a twin, JFK staged his own death, and the moon landing was both real AND fake—they launched twice, once for the cameras, once for the ritual—

The Queen of England isn't who you think—they swapped her in 1998, the real one ran away during the Bird Flu test run—YES it was a test—YES it worked—YES they know how to steer birds into targeted migration paths now—

Your teachers were part of it—not all, some were real, some were lost—but your math teacher? The one who always wore green? That was a psychoacoustic gatekeeper—

Do you drink tap water? WHY? Lithium traces. Microplastic neurons. Thought dampeners. I drank it for years—I learned to smile like they taught me—I passed all the tests except the last one, the empathy inversion, I FAILED and that's why I'm here—

They read minds through earbuds—don't wear them—don't even carry them—they pulse your brain into compliance curves—

And oh God—the assassinations—every celebrity who "died young"? They knew too much. Heath Ledger, Brittany Murphy, the others, the silent ones—their eyes were too open, and the world doesn't like open eyes—

There's a black site behind Disneyland—I got close—I
saw Goofy take off his head and speak in coordinates—
DO YOU BELIEVE ME YET?

Because the boy in front of me looks like a statue, like
disbelief incarnate, like someone who still thinks
breakfast cereal is just breakfast cereal—it's not—it's
how they encode slogans into your muscle memory—

They took me—yes THEY—mind wiped me through
lullabies and rewound films—they showed me war as
advertisement, made me reenact lies with my eyes
open—

I escaped.

I clawed my way out of the back of a moving truck full of
counterfeit organs and half-remembered chants from
dead dictators—

Please.

Just drive.

Before they reroute your memories. Before they tell you I was never here. Before your name gets overwritten like mine did, three times before Tuesday.

I swear to you—this world is stitched together with betrayal—and I'm coming loose.

Help me before I unravel.

Prompted in the style of someone suffering from tachyphrasia, write a poem about all the conspiracy theories that make people feel unsafe and unable to do anything to make things better. try to cover every possible conspiracy theory. Maybe it's a homeless. looking man trying to tell a person waiting for a bus what's real in his world.

Details

event

July 12 at 9:23 PM

apps

Gemini Apps

chat

Hey, hey you, yeah, you there, waiting for the bus, you gotta listen, gotta hear it, all of it, it's everywhere, they're everywhere, can't escape it!

The moon landing, fake, right? Green screen, probably, And aliens, oh they're here, walking among us, Reptilians, shapeshifters, pulling strings, Controlling everything, everything, you see? Fluoride in the water, mind control, that's what it is, Turning us into sheep, docile, numb, Can't think for ourselves, can't fight back, They want us weak, they want us quiet.

The deep state, shadow government, Meeting in secret, plotting our downfall, They caused the market crash, they started the wars, For profit, for power, always for power. Chemtrails in the sky, spraying us with chemicals, Making us sick, making us dumb, Depopulation agenda, that's their game, Thinning the herd, that's what they call it.

Big Pharma, cures hidden, they got 'em, But they won't share, no, no, no, Gotta keep us sick, gotta keep us buying, Their pills, their potions, their poison. And vaccines, don't even get me started, Microchips, tracking us, every move, every breath, They know where you are, they know what you think, They're listening, always listening.

Flat earth, yeah, that's it, it's not a globe, They lied to us, all of them, the scientists, The schools, the news, all lies, all smoke and mirrors, To keep us in the dark, to keep us small. And the internet, oh, it's a trap, Data mining, selling your soul, They know your secrets, your fears, your desires, They'll use it against you, they always do.

JFK, still a mystery, right? Who really did it? Was it the mob? The CIA? The aliens again? They never tell us the truth, never, Always a cover-up, always a lie. And pizzagate, you heard of that? Child trafficking, Underground tunnels, hidden horrors, The elite, they're all involved, Monsters in suits, smiling faces, but monsters.

So you see, you see it now, don't you? It's all connected, all part of the big plan, To keep us down, to keep us scared, To keep us from ever changing anything. We're just pawns, man, just pawns in their game, Can't do anything, can't fight back, It's hopeless, you hear me? Hopeless! Now, you got a spare dollar? For a cup of coffee? Just one?

Why this activity?This activity was saved to your Google Account because your Gemini Apps Activity setting was on while using Gemini Apps.info_filledIf you use a shared device or sign in with multiple accounts, activity from another account may appear here. Learn morei want a poem of several minutes in length on the subject and in the style of someone suffering from tachyphrasiaEditI'll create a poem that captures the rapid, stream-of-consciousness flow characteristic of tachyphrasia - that compulsive, pressured speech pattern where thoughts tumble out in an unstoppable cascade.Rush of Words - A Tachyphrasia PoemDocument

This poem captures the breathless, unstoppable quality of tachyphrasia - the way thoughts and words tumble out in an endless stream without the usual pauses and breaks. The structure mirrors the condition itself: no line breaks, minimal punctuation, thoughts that spiral and loop back on themselves, and a desperate quality of someone trying to keep up with their own racing mind.

The poem explores themes often associated with pressured speech: the fear of silence, the compulsion to fill every gap with words, the feeling of being carried along by an unstoppable current of language, and the paradoxical search for peace within the chaos of constant expression.i want a poem of several minutes in length on the subject of inside knowledge about assasinations and from the point of view of someone who has just escaped being kidnapped and is asking for help and in the style of someone suffering from tachyphrasiaEditI'll create a poem that captures the frantic, pressured speech of someone who has escaped kidnapping and has urgent information about assassination plots. This will maintain the tachyphrasia style while focusing on the psychological state of someone in extreme distress trying to communicate vital information.Testimony in the Dark - An EscapeDocument This poem captures the frantic, breathless testimony of someone who has escaped captivity with crucial information about assassination plots.

The tachyphrasia style emphasizes the psychological state of someone in extreme distress - the words pouring out without pause, thoughts spiraling and looping back on themselves, the desperate need to be believed and to communicate vital information before it's too late.The poem explores themes of trauma, urgency, paranoia, and the terrible burden of dangerous knowledge, all filtered through the lens of someone whose normal speech patterns have been disrupted by extreme stress and the compulsive need to share what they've learned before more harm occurs.sestina with the title "something sad about serendipity"

.EditSomething Sad About SerendipityDocument I've written a sestina titled "Something Sad About Serendipity" that follows the traditional form's complex structure. A sestina consists of six stanzas of six lines each, plus a three-line envoi (conclusion). The six end-words (see, chance, fate, doors, story, find) rotate in a specific pattern throughout the poem, creating a intricate weaving effect that mirrors the theme of serendipity itself.The poem explores the melancholy paradox of serendipity—how the more we search for meaningful coincidences and connections, the more we might miss them when they actually occur. It touches on themes of missed opportunities, the timing of love and fate, and the bittersweet nature of looking back on the chances we didn't take or recognize.j

Rush of Words

Words rushing rushing like water breaking through dam walls can't stop them coming faster than I can think faster than I should speak but here they come anyway tumbling over each other like children down a hill rolling rolling rolling and I'm trying to catch them all but they slip through my fingers like mercury like time like the very thoughts that birth them before they're fully formed before they're ready to be born but out they come anyway ready or not here they come ready or not here I am speaking speaking speaking without pause without breath without the luxury of silence that normal people take for granted the spaces between words the comfortable pauses the thoughtful hesitations that I've forgotten how to make forgotten how to be forgotten how to simply be quiet and let the world exist without my constant commentary my running narration of everything I see feel think remember anticipate worry about celebrate mourn for laugh at cry over and did I mention I'm laughing now or was that crying it's hard to tell when the emotions come as fast as the words when everything blends together in this river of sound this torrent of syllables this avalanche of language that started somewhere back there in the mountains of my mind and now it's picked up speed picked up debris picked up pieces of every conversation I've ever had every book I've ever read every song I've ever heard every thought I've ever thought and they're all mixed up together like ingredients in a blender like colors on a palette like the inside of my head right now which is spinning spinning spinning like a top like a planet like the whole universe expanding faster than light faster than sound faster than I can keep up with but I have to keep up I have to keep talking because if I stop if I pause if I take that breath I'm supposed to take the silence might swallow me whole might reveal the emptiness underneath might show that all this rushing water is just covering up the drought the terrible drought of having nothing real to say nothing important to contribute nothing worth this flood of words this deluge of discourse this tsunami of syntax that I can't turn off can't slow down can't control any more than I can control the weather or the tides or the way my heart beats beats beats against my ribs like a bird in a cage like words in my throat like thoughts in my skull demanding release demanding expression demanding to be heard even if they're not worth hearing even if they're just noise just sound just the desperate attempt of a mind that's forgotten how to rest how to pause how to let the world exist in peace without my constant need to fill every silence with sound every gap with words every moment with meaning even when there is no meaning even when the words have lost all connection to the thoughts that spawned them have become just reflexes just habits just the way water flows downhill without thinking about it without choosing to without any ability to stop until it reaches the sea and maybe that's what this is maybe this is my sea this endless talking this ceaseless speaking this perpetual flood of language that pours from me like wine from a broken bottle like light from a dying star like the last words of someone who's been holding their breath too long and finally finally finally has to let it all out in one great rushing exhalation that carries with it every unsaid thing every swallowed word every suppressed thought every bitten tongue every polite silence that I've endured and now I'm making up for lost time making up for all those moments when I should have spoken but didn't when I should have said something but stayed quiet when I should have filled the air with words but let others do the talking and now it's my turn now it's my time now it's my moment to speak and speak and speak until the words run out until the thoughts dry up until the very language itself begs for mercy begs for rest begs for the simple gift of silence that I can't seem to give that I've forgotten how to offer that feels like death like emptiness like the end of everything I am because if I'm not talking then who am I if I'm not speaking then what am I if I'm not filling the air with sound then how do I know I exist how do I know I matter how do I know I'm here at all and that's the fear isn't it that's the terror that drives this engine of expression this machine of meaning this factory of phrases that works overtime every time producing more than anyone could ever need or want or use but still I produce still I create still I manufacture these moments of sound these fragments of thought these pieces of myself scattered like breadcrumbs through the forest of conversation hoping someone will follow them back to where I started hoping someone will understand the path I've taken the journey I'm on the destination I'm trying to reach through all these words all these sounds all these attempts to bridge the gap between what I think and what I say between what I feel and what I express between who I am and who I appear to be when I'm talking like this when I'm racing like this when I'm rushing like this toward some finish line I can't see some goal I can't name some ending I can't imagine because the words keep coming keep flowing keep rushing like a river that's forgotten the ocean exists like a clock that's forgotten how to stop like a heart that's forgotten how to rest between beats and maybe that's what this is maybe this is just the rhythm of a mind that's learned to race instead of walk learned to run instead of stroll learned to sprint through every thought every idea every observation every memory every dream every fear every hope every wish every regret every joy every sorrow every moment of being alive and conscious and aware in this world that moves so fast that talks so loud that thinks so much that I have to talk faster think louder move quicker just to keep up just to stay relevant just to remain part of the conversation that everyone else seems to be having at a normal pace with normal pauses with normal breath with normal silence between the words between the thoughts between the moments of being human and maybe that's what I'm trying to find in all these words maybe that's what I'm searching for in all this sound maybe that's what I'm hoping to discover in all this rushing water the simple quiet human moment the peaceful pause the restful silence that exists somewhere beyond the flood somewhere after the storm somewhere past the place where words end and being begins but I can't find it can't reach it can't stop long enough to look for it because the words keep coming keep flowing keep rushing like water like time like life itself moving faster than I can follow faster than I can understand faster than I can control carrying me along in their current carrying me toward some destination I can't see some shore I can't imagine some place where the talking stops and the living begins where the speaking ends and the being starts where the flood of words finally finds its ocean and becomes still becomes quiet becomes the silence I've forgotten how to make forgotten how to be forgotten how to find in the spaces between the words in the pauses between the thoughts in the rest between the beats of this heart that won't stop racing this mind that won't stop talking this voice that won't stop speaking until it remembers how to whisper how to murmur how to simply be quiet and let the world exist without commentary without narration without the constant need to fill every silence with sound every gap with words every moment with meaning that may or may not be there may or may not matter may or may not be worth all this effort all this energy all this desperate attempt to be heard to be understood to be known through the medium of language through the vehicle of voice through the vessel of words that carry us from thought to thought from moment to moment from breath to breath until finally finally finally we find the courage to stop to pause to rest in the silence that waits patiently beyond the last word beyond the final sound beyond the end of this endless beginning that we call speech that we call expression that we call the need to be heard in a world that may or may not be listening may or may not be there may or may not care about the words we speak the thoughts we think the lives we live in the rushing water of language that carries us all toward the sea toward the silence toward the place where words end and something else begins something quieter something deeper something truer than all the words we use to try to capture it to try to contain it to try to make it ours but it can't be captured can't be contained can't be owned by words by thoughts by the endless stream of consciousness that flows through us like water through a pipe like light through a window like time through the hours of our days and nights and years and lives that pass so quickly that we have to talk fast just to catch them have to speak quickly just to keep up have to rush through every word every thought every moment just to stay alive just to stay present just to stay here in this rushing water this flowing time this river of words that carries us all toward the same destination toward the same silence toward the same rest that waits for all of us beyond the last word beyond the final sound beyond the end of this poem that I'm writing now that I'm speaking now that I'm thinking now faster than I can understand faster than I can follow faster than I can control but maybe that's okay maybe that's right maybe that's how it's supposed to be this rushing water this flowing time this river of words that carries us home.

Testimony in the Dark

Please listen please you have to listen I escaped I got away but they're still out there still planning still moving forward with what they told me what they said in that room that cold concrete room where they kept me for days or weeks I don't know time stopped meaning anything when you're tied to a chair when you're blindfolded when you're listening to voices in the dark discussing things that shouldn't be discussed planning things that shouldn't be planned and I heard everything I heard names I heard dates I heard locations and they thought I couldn't understand thought I was just cargo just leverage just something to be moved from place to place but I was listening always listening because what else can you do when you're helpless when you're afraid when you're hoping someone will come for you but knowing they probably won't and the voices kept talking kept planning kept organizing and I memorized every word every name every detail because I thought if I survive if I get out if I make it through this I have to tell someone I have to warn someone I have to stop what they're going to do and now I'm here I'm free I'm alive but barely barely holding on because the adrenaline is wearing off and the fear is setting in the real fear the deep fear that comes after the immediate danger has passed when you realize what you know what you've heard what you have to do and I don't know who to trust I don't know who's safe I don't know who might be part of it because they mentioned names familiar names important names people in government people in business people who should be protecting us but maybe aren't maybe they're part of the plan maybe they're the ones who need to be warned or maybe they're the ones doing the warning to the wrong people and I'm spinning I'm spiraling I'm trying to organize the information in my head but it's all jumbled all mixed up all flowing together like water like blood like the sweat that's pouring down my face right now because I'm terrified absolutely terrified that I'm too late that it's already started that while I was escaping while I was running while I was hiding they were moving forward with their timeline with their schedule with their plan to eliminate targets to remove obstacles to clear the path for whatever comes next and I heard them say September I heard them say the courthouse I heard them say the hotel I heard them say the names of people who matter people who make decisions people who could change everything if they live long enough to do it but they won't if I don't speak up if I don't tell someone if I don't find the right person to trust with this information this terrible knowledge this burden that I never asked for never wanted never imagined I'd have to carry but here I am carrying it like a weight like a stone like a bomb that might explode at any moment and take me with it take everyone with it because that's what secrets do especially dangerous secrets especially secrets about death about murder about plans that should never be made should never be spoken should never be carried out but they will be they will be unless someone stops them unless someone acts unless someone believes me believes what I'm telling you believes that I'm not crazy not paranoid not making this up but telling the truth the terrible truth that I learned in that room that dark room where they kept me where they talked where they planned where they organized the end of everything good everything decent everything that stands between us and chaos and I know how this sounds I know how I sound frantic desperate unhinged but wouldn't you be wouldn't anyone be if they'd heard what I heard if they'd seen what I saw if they'd been where I was for all those days or weeks listening to the casual discussion of murder listening to the methodical planning of assassination listening to the calm professional voices organizing the destruction of people who don't even know they're in danger don't even know they're targets don't even know that somewhere in a concrete room men with clipboards and cell phones are deciding when they die how they die who will do it who will cover it up who will benefit from their absence from their silence from their permanent removal from the equation and I can't stop thinking about it can't stop hearing those voices can't stop seeing those faces in the shadows because even though they kept me blindfolded I caught glimpses I saw fragments I observed details that I shouldn't have seen shouldn't have remembered shouldn't be able to describe but I can I can because fear sharpens everything makes everything clearer makes everything more vivid more real more impossible to forget and I remember the scar on the hand that held the phone remember the accent that pronounced certain words a certain way remember the cologne that one of them wore remember the sound of their shoes on the concrete remember the way they said goodbye to each other like they were going to work like they were heading to the office like they were normal people living normal lives instead of monsters planning monstrous things and now I'm out I'm free I'm alive but for how long how long before they realize I'm gone realize I escaped realize I might talk might tell might expose their plan and come looking for me come hunting for me come to finish what they started because I know too much have heard too much have seen too much to be allowed to live to be allowed to speak to be allowed to warn the people who need to be warned and that's why I'm here that's why I'm talking that's why I'm begging you to listen to believe to act because time is running out time is almost up time is about to expire and when it does when the moment comes when the plan unfolds those people those targets those names I heard in the dark will be gone will be dead will be silenced forever and it will be too late too late to stop it too late to prevent it too late to save them unless you help me unless you listen unless you believe that what I'm telling you is real is true is happening right now while we sit here while we talk while we waste precious seconds that could be used to warn them to protect them to stop the machinery of death that's already in motion already moving already approaching its destination and I know the date I know the time I know the place I know the method I know the backup plans I know the contingencies I know the escape routes I know the cover stories I know everything because I was there I was present I was witness to the planning of murder disguised as accidents disguised as heart attacks disguised as suicides disguised as random violence but none of it random none of it accidental none of it natural all of it planned all of it organized all of it designed to look like anything but what it really is which is assassination systematic elimination methodical removal of obstacles of problems of people who stand in the way of something bigger something darker something that I only partially understand but understand enough to know that it's wrong that it's evil that it has to be stopped and I'm the only one who can stop it the only one who knows the only one who heard the only one who escaped the only one who can warn them can save them can prevent the cascade of death that's about to begin unless you help me unless you act unless you pick up that phone and make the calls that need to be made send the messages that need to be sent alert the people who need to be alerted because I can't do it alone I can't carry this burden alone I can't fight this fight alone and I'm running out of time running out of strength running out of hope that anyone will believe me that anyone will listen that anyone will care enough to act on what I'm telling you and maybe that's their plan maybe that's what they're counting on that even if someone escapes even if someone talks even if someone tries to warn the targets no one will believe them no one will listen no one will take action because it sounds too crazy too impossible too much like a movie like a story like something that doesn't happen in real life but it does happen it is happening it's happening right now and I'm the proof I'm the witness I'm the survivor who got away who lived to tell who made it out of that room that prison that place where they plan the end of the world one death at a time one murder at a time one assassination at a time and I have to stop it have to prevent it have to save them even if it costs me everything even if it puts me in danger even if it makes me a target too because that's what you do when you know what I know when you've heard what I've heard when you've seen what I've seen you act you speak you fight even when you're scared even when you're shaking even when you're talking so fast that the words tumble over each other like water like blood like time running out running away running toward the moment when it will be too late when the plan will unfold when the targets will fall when the world will change forever and all because I couldn't make someone listen couldn't make someone believe couldn't make someone act on the information the knowledge the terrible truth that I carry like a weight like a curse like a responsibility that I never asked for but can't abandon can't ignore can't walk away from because they're counting on me the targets the victims the people who don't know they're about to die are counting on me to save them to warn them to stop the machinery of death that's grinding toward them like a train like a storm like fate itself and I won't let them down I can't let them down I have to find a way to make you understand to make you believe to make you act before it's too late before the clock runs out before the plan unfolds before the assassinations begin and the world changes forever.

Made in the USA
Monee, IL
04 November 2026

67387910R00026